On the Move
Advanced Individual Training

Vickey Joins the Military Series

Book 3

On the Move
Advanced Individual Training

Book Three

Table of Contents

Advanced Individual Training

What is the Job?

What a relief it was, having completed the toughest part of my journey to becoming a soldier: basic training. It marked a significant milestone in my military path, and I survived.

I was then on the second leg of my journey to becoming a soldier, arriving at my advanced individual training, commonly known as AIT. AIT was where I would learn my military occupational specialty or the job I would do in the military.

Around midnight, myself and approximately thirty others arrived at the training facility. We were promptly directed to a designated area for rest, which, as it turned out, was yet another holding station.

Morning broke, and I, along with two other female recruits, stirred from sleep. We were instructed that formation was scheduled for 5:55 am. “Oh not more early morning work starts,” I thought to myself.

Since we arrived over the weekend, our time was filled with sleeping, eating, and picking up trash around the facility.

“Your stay here will only be as hard as you make it,” the sergeant explained.

“If you do what you are told when you are told to do it, everything will be easy,” the sergeant continued.

I was sure that the laid-back atmosphere that we were experiencing would end on Monday morning. I could not help but reflect on the transition from the reception center to basic training.

“Shake it off Vickey,” I thought as I shook my head trying to erase that memory.

The lights were out at 8:00 pm. I grabbed my headset, turned out the lights, and listened to music as I drifted off to sleep, but I could not help but wonder what tomorrow would bring.

We woke the next morning to more sleeping, eating, trash collection, and just general free time.

“I’m not sure if I should be worried about what to expect tomorrow as we transition from the holding facility to our actual training site,” I kept thinking about the move.

“Oh, please do not be another ‘shock and awe’ experience, I am not sure I can endure another one of those,” I thought.

Well, the day had arrived for us to transition and process into our advanced individual training, again referred to as AIT.

AIT

The Early Days

With our early arrival at training, we found ourselves facing about two weeks of waiting before the actual training kicked off. Nonetheless, the desert climate offered a far more favorable environment compared to the conditions I endured during basic training.

I had started forming a bond with one of the other new trainees, and we made a pact to stick together and support each other through the initial days of training. AIT did not appear to be anything like basic training.

It was as if the trainees were never exposed to basic training because their marching skills were extremely poor, they were overly talkative, and seemed like total screw-ups and very immature. The stark difference from what I had experienced in basic training made the first day at the new training station rather disheartening.

As the days went on, leading up to training, we did chores, ate, and attended briefings. With each passing day, the initial unease subsided, and we grew more accustomed to the routine. We were told that things would get better, and I believed it.

After a few days at the new training station, we took our initial PT test to gauge our current fitness levels. My score had improved from what I achieved during basic training.

"You surprised yourself, ha," the drill sergeant said.

"With all your education, you ought to know by now, you can do anything you put your mind to," he said.

One week into the AIT hold status, tensions escalated between the United States (US) and the Middle East; the US was on the eve of a deadline for war to begin. Despite the looming specter of war, we received instructions that training would proceed as scheduled until further notice, even though we had yet to commence actual training. At that given moment in time, I was not afraid.

The days leading up to training were torturous. Awaken at "oh-dark-thirty" to do excruciating physical activity. One constant was that we could count on someone not doing what they were supposed to do and that always, always resulted in us being hauled outside in the early morning to be brutalized.

In the days leading up to the start of actual training, the crucial deadline set by the US to Saddam Hussein to get out of Kuwait had arrived. However, on that fateful day, there was more unwelcomed news in store for the hopeful individuals and impending graduates who anticipated a reprieve or time off from training. They learned that all leave had been abruptly canceled, and a freeze was enforced, effectively grounding everyone with no option for departure. They all had to remain at their current duty location until otherwise notified.

As the deadline day passed without any significant developments on the war front, it brought a sense of relief. The day began with thc usual early morning routine and its accompanying torture. However, by late afternoon, a special news report broke the news that war had indeed begun in Iraq.

As we watched the report, facial expressions transitioned from curiosity to tension, reflecting the gravity of the situation. It was the moment when reality truly sank in: upon completion of our training, we could be sent to war.

The then-President, George H. W. Bush, spoke to the nation to inform that he had signed the forms necessary to begin war and to let everyone know that the US had made its first attack at 3:00 am Iraq time.

Everyone was confined to the post, even the cadre, which included drill sergeants and commanders. Things became very tense; suspense and anxiety rose. All passes were canceled, and anyone who was on pass before the fighting began was brought back to the post. No one could leave the post for any reason.

After the announcement and everyone returning to post, there wasn't much else that changed. However, they did close off several entry points and increase security measures, but training continued as usual.

The Days of Training

Finally, training began —the training to learn the Army job we would perform. The job, we learned, was working with the air defense system that was doing well in the war. That realization brought about some anxiety, as it increased the likelihood of being deployed to the war zone after training concluded.

The training days began as regular school days, with classroom instruction and out into the field where we would put that training into practical use.

But on a Tuesday, during the early days of training, there was a request to film our training, and it was approved. That was exciting news.

Another significant day unfolded during our training when the Sergeant Major of the Army, the highest-ranking non-commissioned officer in the Army, paid a visit to address the class. There was a sense of uncertainty regarding the reason for the heightened attention, though it likely stemmed from the success of the weapon system we were training on and its success in the ongoing war.

“Platoon guide, Specialist Robinson,” the Sergeant Major of the Army said as he began his address to me.

Being platoon guide was a leadership position, and with the position came some privileges, such as getting your own room and not having to sleep in the bay area with twenty or more personnel. But also, it came with the opportunity to speak to high-ranking leaders.

Trying to make the room homey!

"Yes, Sergeant Major," I responded, and the conversation went on for a brief period. The excitement over having spoken with the Sergeant Major was the highlight of the day. The drill sergeants expressed that getting an opportunity to speak with the Sergeant Major directly was a privilege because no one ever got to meet him, particularly trainees like us.

Wow, how do you top that day, meeting the Sergeant Major of the Army, the top non-commissioned officer; you do not. All other days were routine, until.

On another school day, a day expected to be like all others, it was announced that the Vice President of the United States, Dan Quayle, would be coming to speak with the class. We were all lined up and within feet of shaking the hand of the Vice President, and then it happened.

"Oh, my goodness," I thought to myself.

"How many people can say they shook the hand of the Vice President of the United States? I can."

"You guys are pretty darn lucky," said one of the sergeants.

"Because in three weeks, you have had an opportunity to meet the Sergeant Major of the Army and the Vice President of the United

States. This is something that most people who have been in the Army for years haven't had an opportunity to do."

As we came off our good fortune, we returned to our uneventful training days, early morning PT, barracks chores, breakfast, and off to class. Even though I had my own room, keeping it inspection-ready wasn't straightforward, especially when my schedule was controlled by others. All my belongings were stored in a wall locker.

As the weeks ticked down, a month after the war had started in the Middle East, the commander announced that Saddam Hussein had agreed to negotiate with the United Nations to sign an agreement that he would clear out of Kuwait. That announcement was welcomed news.

Hearing that the immediate war had ended, we were able to feel relieved. That was a bit of assurance that we would not be heading to a war zone any time soon. But we were soon informed that although the immediate fighting had ended, there were still American troops in that region, and there was still a strong possibility that we could end up in Kuwait. Our emotions took a bit of a dive, but again, there was no fear. Signing up for the military, there was always that possibility of having to go to war.

With week five of training looming, we wrapped up our first four weeks of training by taking final exams, practical and written, as we prepared to move on to the equipment side of training, where we would spend six weeks learning the particulars of the air defense equipment.

Learning to drive the 10-ton vehicles!

But before the transition, of course, there had to be multiple drill sergeant shake-downs, unannounced barracks checks, and many chores had to be completed before we had our big transition inspection.

We dedicated a day off from school to prepare diligently. We invested much of our time in arranging our wall lockers, displaying our shoes, and placing our brass on our dress uniform. Despite our efforts, we failed the inspection, leaving us bewildered as we had believed we had prepared adequately. Honestly, I thought it was all just a mind game; I doubted there was ever any intention of them allowing us to pass that inspection.

Trainee Time Rolling On

To break us out of our routine, the drill sergeants organized a sport-a-thon day. It provided a welcomed break from our usual activities and turned out to be a fun-filled day for all.

As the days passed, the monotony of military life became increasingly dull. Aside from everyone living in the same place, dressing alike, being told when you could eat and sleep, and being awakened early just because, military life or the life of a soldier was turning out to be quite routine.

With several more weeks remaining before training was expected to end, I had begun to grow tired of the entire masspunishment ordeal and the routineness of it all. I was no longer buying into the whole "together they fall" concept. I had had enough.

Weekends during AIT weren't too bad. Since we were in our advanced training, we found that getting time off away from the training site was a bit more common. I believed it was more so for the drill sergeant's peace of mind and less about doing something for the trainees.

And with some of that time on the weekend, our class decided to give the upcoming graduating class something to remember about their AIT days. With days before their graduation, we set out on "Operation Sendoff."

In the late hour, about 11 pm, myself and several others dressed in mission gear and headed out to commence "Operation Sendoff." Just as we arrived at the mission site (male barracks), one of the drill sergeants was driving by.

"Who is that?"

"What are you doing?"

We went to the vehicle and explained what we were planning to do, give the graduating class a parting gift. Surprisingly, the drill sergeant said…

"Okay, have fun," and smiled and left.

Once the drill sergeant left, we resumed our mission. By around 1:30 am, we were apprehended and summoned to the night sergeant's office. After a short lecture, we were allowed to go to bed.

Thinking we had gotten away with everything, we soon discovered otherwise. Later that day, we were summoned to the drill sergeant's office, where he informed us that he would let us know our punishment later. Well, we were handed a shocking surprise. We were all released on pass.

"Get the h--- off the hill," the drill sergeant yelled.

Off we all went, as quickly as we could get changed, and left the training area. We looked forward to the passes, the time away from the training site. The passes were small doses of freedom and liberty.

The experience served as a stark reminder of how the freedom to move about as we pleased was often taken for granted until it was suddenly restricted. It was the small, simple things—the ability to move freely—that were sorely missed during training.

Training continued and the days went on with a great deal of routine, physical training, and classes.

Many, when they hear the word military, conjure up visions of bombs, firing weapons, heavy military equipment, and constant fighting. The reality of the military, yes, has all those things, but an overwhelming part of the military is as routine as any other job; that came as a shock. Weapons, fighting, and mud trekking were rare.

As testing was wrapping up on the tactical aspect of the job, information regarding orders and location of the first permanent party duty stations was announced. Most of the class was headed to Germany.

Many of us knew that if we were heading to Germany, there was a strong likelihood that we would also travel to the Middle East region because the US was still conducting missions there.

But training was not over yet. As the end of training grew closer, that entailed more intensive training and testing, and practical training exercises, putting the book knowledge to practical use.

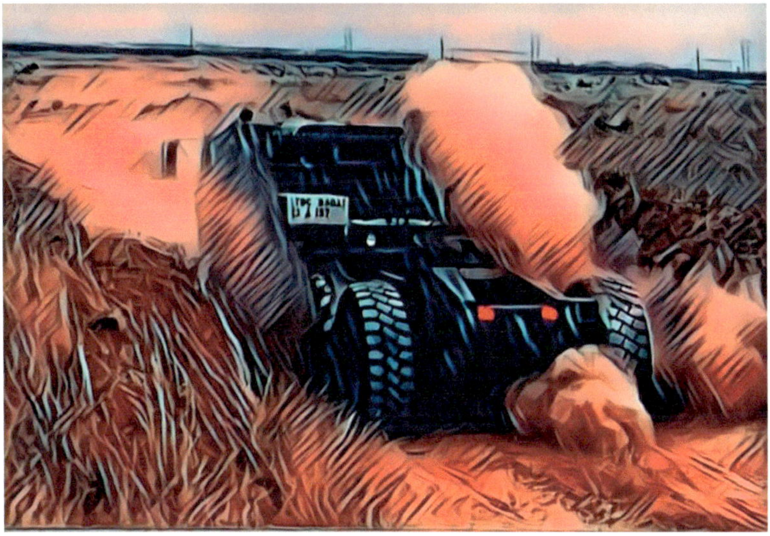

Graduation day was approaching fast, and preparations for many great ceremonies were underway. It was only a matter of days before I would officially become a full-fledged soldier.

The look of the final days of training!

The day finally arrived for graduation. In the early days, it appeared as though the day would never arrive, but it had, and there was no real thrill. Yes, there was the relief of having it over, but no big thrill came over me. I thought it would come later during the ceremonies.

Graduation!

As the ceremonies concluded, I found myself surprisingly devoid of any overwhelming thrill. Now, the moment had arrived to bid farewell to my second training site and transition to my first permanent duty station.

Hanging on the quad as I await departure!

After some time reflecting on the training that I had gone through, both basic and AIT, it became clear that there was a reason for all we went through.

The military with all its rules and regulations, to a civilian, may seem controlling or unnecessary; but it later became clear that all the unrelated, mundane acts that we went through could mean life or death for a soldier in combat. The act of making a bed a certain way, marching in unison, wearing the uniform a certain way, or staying in cramped living quarters may have initially seemed mundane or a waste of time, but were necessary to teach us civilians discipline and teamwork, both, the bedrock of military service.

I continued to reflect on my journey during that period, especially considering I had three weeks before reporting to my first permanent duty station in Germany after AIT. I took advantage of the gap to travel home and unwind.

Returning home, I felt like a different person, having undergone experiences that none of my acquaintances had encountered. I had overcome fears and challenges I had never even considered before. Now, the possibilities ahead seemed limitless.

I had dreamed of travelling and now my next big, unexpected move would be out of the country. I enjoyed home, and its familiarity, but the pull of future adventures was strong.

I woke up around 4:30 am to get ready for the long day ahead, as I was set to travel to Germany, an eight-and-a-half-hour flight away.

It finally sunk in - I was in the Army, a fully trained soldier ready to fulfill an important role. The destination ahead promised to be significant, with my duties taking center stage.

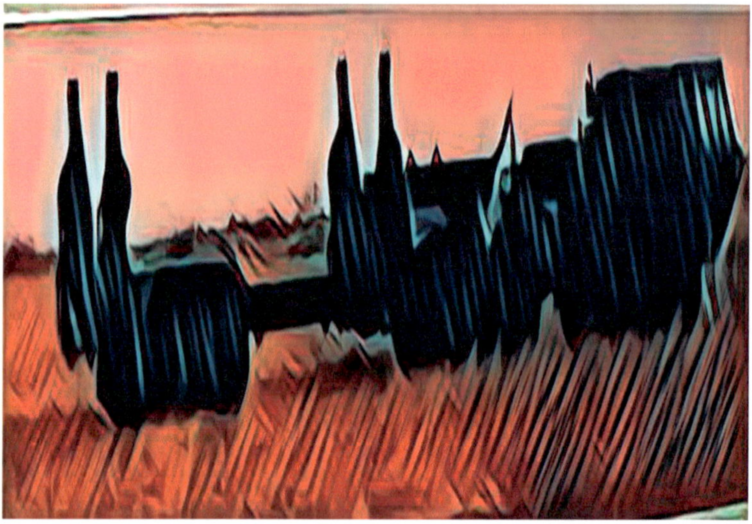

The flight arrived in Germany at midnight, state-side time, but it was 7:00 am in Germany, and that meant despite an exhaustive two days of travel, we hit the ground and went straight to work.

Welcome to your first permanent duty station, Frankfurt, Germany; you are not in the US anymore.

And I was no longer a civilian. I was now a soldier in the United States Army, and that alone would bring with it a complete set of adventures.

Made in the USA
Las Vegas, NV
07 November 2026

57669929R00021